THE VICTORIAN ARMY IN PHOTOGRAPHS

By the same author
The Zulu War

THE VICTORIAN ARMY IN PHOTOGRAPHS

DAVID CLAMMER

DAVID AND CHARLES
NEWTON ABBOT LONDON VANCOUVER
NORTH POMFRET (VT)

0 7153 6853 2

Set in Imprint and printed in
Great Britain by Redwood Burn Limited
Trowbridge and Esher
for David & Charles (Holdings) Limited
South Devon House Newton Abbot Devon

Published in the United States of America
by David & Charles Inc.
North Pomfret Vermont 05053 USA

Published in Canada
by Douglas David & Charles Limited
132 Philip Avenue North Vancouver BC

CONTENTS

Introduction	6
People	10
Equipment and Uniforms	15
Training	36
Transport and Communications	45
Recreation	57
Campaigns	66
Sources and Acknowledgements	95
Index	96

INTRODUCTION

The Victorian period was a busy time for the army; the demands of empire were many, various and frequent. The *Pax Britannica* was maintained only by incessant small wars, campaigns and punitive expeditions, of which, during the sixty-four-year reign, over 200 were fought at an average rate of about four a year. With the exception of the Crimea and the two Boer Wars, this unremitting fighting took the form of 'native' wars, which fact imparted to the army its character and quality.

After the Napoleonic Wars, Britain, secure in her naval supremacy, abstained from direct intervention in European affairs for almost a century, and at a time when the major powers of the continent were raising mass conscript armies, maintained a small volunteer army scattered across the world to protect, police and expand the territories of the empire. In Europe a new concept of the Nation in Arms was formed from the lessons of the American Civil War and the Franco-Prussian War. The significance of population, industrial potential and strategic railways was realised: general staffs, trenches and barbed wire were developed with their accompanying technical and intellectual skills, presaging the advent of modern total war. All these things were largely ignored in Britain as irrelevant to the colonial experience.

One result of this at the practical level was the retention (until the shock of the Boer War) of outmoded tactics. As late as 1898, in the Sudan, the cavalry charge and the infantry square, the bayonet and volley firing, were employed precisely as they had been at Waterloo, while in Europe such methods had long been rendered obsolete, not to say suicidal, by improvements in both artillery and small-arms.

Besides superior fire-power, the keys to success against savage opponents were personal courage, discipline and *esprit de corps*. Fighting was very much a regimental business, and it was regimental qualities that were sought and fostered. Conditions of service, the small scale of most wars, and the prolonged exile in remote foreign stations made this inevitable. It could be said indeed, that an army as such hardly existed at all, but rather a collection of individual regiments thrown together by the exigencies of war, and which, until the abolition of purchase, belonged as much to their colonels as to the nation. Such a system, while it had its successes, was not without its drawbacks: the approach was too often amateur rather than professional; improvisation tended to replace organization; personalities were substituted for systems.

The Victorian period was indeed the great age of the popular military hero: of Havelock struggling to the relief of Lucknow; Gordon dying his martyr's death at Khartoum; and later the avenging Kitchener rolling back the tide of Madhist barbarism; Roberts marching on Khandahar through the mountain fastness of Afghanistan; Baden-Powell beleaguered in Mafeking; and Wolseley of course, 'Our only General', performing brilliant feats the length and breadth of the Empire. These few epitomised for the Englishman an image of his better self, courageously determined to propagate his stern ideals of duty, justice and Christian civilization, and were commemorated not only publicly and officially, but by the ordinary citizen, in popular expressions, pictures and household objects.

Yet heroes are men apart, and during the early part of the period the army in general was the object neither of interest nor respect, except on the part of the queen, whose concern for military affairs was keen and unflagging. The empire was not particularly popular, and was regarded as a liability rather than an asset. The soldier, in consequence, was the object of politicians' parsimony and public indifference, and sometimes of hostility. After about 1870, however, interest in the imperial idea quickened, as the race with other European powers for overseas territories intensified, and the soldier, in the vanguard of this scramble for empire, became the subject of patriotic – some would have said jingoistic – pride and popularity.

But in whatever esteem the army was popularly held, it remained as an institution socially isolated and unrepresentative of the nation as a whole. The officers were drawn exclusively from the upper, or at least moneyed class, while the rank and file came in the main from the poorest and lowest levels of society. Both joined the colours for compelling social reasons: the one because their status and traditions left them few acceptable alternatives; the other because hunger and destitution often left them none. The growing middle class, who stood to gain most on the rising tide of national power and prosperity, never shared the military burden of empire. This, perhaps, is not surprising, for most of the little wars of the reign took place for reasons too obscure, in places too remote, and against foes too outlandish to interest the average Englishman.

The army at the time of Victoria's accession was still in essence the army of Waterloo; by the end of the period it was a recognisably modern force. The transformation was always slow and often reluctant. Sometimes the process was one of natural evolution. Sometimes changes were wrought by the salutary, if painful, agency of outside events such as the Crimean fiasco, or Cardwell's reforms, or the humiliations of the Boer War. The commander-in-chief at last became subordinated to the secretary of state for war, and for the first time the army passed under full political control. The purchase of commissions was abolished and the professional training of officers improved. Conditions of service in the ranks were mitigated from barbarism to mere discomfort. Flogging was done away with, and medical facilities, food, pay and accommodation for the private soldier were improved. Long service was replaced by short. The musket gave way to the rifle and machine-gun; breech loading rifled ordnance replaced smooth-bore muzzle loading guns. Signalling improved. And, as symbolic as it was practical, khaki superseded the traditional British scarlet.

An examination of the wars in which the army was so continuously employed gives an overwhelming impression of diversity: of the political diversity of their causes; the geographical diversity of their locations, and the diversity of the opponents they presented.

Sometimes these small wars were self-defensive, as in the case of the Sikh Wars, when British held territory was attacked. Sometimes they were offensive, seeking to advance or protect British interests. The Second China War, for example, was waged in order to enforce trade and diplomatic agreements. Both the wars in Afghanistan were intended to neutralise the supposed intrigues of Russia against India. The Zulu War might be described as an offensive-defensive operation, fought on the principle that the best means of defence is attack, in order to remove a serious threat to a British colony. Occasionally a campaign might be in the nature of an extensive rescue operation: both the Gordon Relief and Chitral expeditions come under this heading. Rebellion sometimes had to be suppressed: the Indian Mutiny was of course the most serious case, while the Red River expedition in Canada also comes to mind. Since the empire was regarded as an instrument of civilization, some wars were undertaken as crusades against barbarism – the Ashanti campaign, for example, or the reconquest of the Sudan. On occasions the army undertook purely punitive expeditions, as so often on the North West Frontier, where imperial prestige was under constant pressure from the unruly Pathan tribes. And sometimes these little wars simply arose out of the inevitable friction occasioned by the contact of white civilization with less advanced peoples, often with European greed for land as the precipitating factor. The wars with the Kaffirs, the Matabele and the Maoris are all instances of this.

In all these multifarious campaigns, superior fire-power and discipline proved a winning combination, and the army was almost invariably victorious, usually with small cost to itself. But it would be wrong to suppose that these victories were easily achieved, for the difficulties of Victorian campaigning cannot be measured by the casualty lists alone.

Conditions of campaigning were rigorous in the extreme. Geography was always a greater obstacle to Victorian generals than hostile savages. They operated in a world not yet fully explored, and in countries of which no reliable maps were available. This even happened in South Africa, in territory that had been under British control for years. Lack of geographical knowledge could result in other difficulties, as it did during the Gordon Relief expedition, when expert opinion on the timing of high Nile, a critical factor, proved to be misinformed.

To problems of topography were added those of climate, which in the then primitive state of medical knowledge might prove more lethal than the enemy. Men could die of fever in the jungles of West Africa, of heat stroke in Egypt and India, or of exposure in the Himalayas, to say nothing of cholera, dysentry, and any number of other exotic conditions. Climate could also impose a time factor on a campaign. Both Wolseley in Ashanti and Napier in Abyssinia had to reach their objectives and withdraw again before the rains made the country to their rear impassable.

Difficulties such as these compounded an already complex transport problem in this premechanised era. Every gun, every round of ammunition, all food, medical supplies, and material and equipment of every sort had to be hauled or carried over every natural obstacle by the vast assortment of horses, mules, camels, oxen and elephants that accompanied every toiling column. On the Abyssinian campaign the ratio of animals to troops was three to one, while on the Tirah expedition no less than 60,000 were required. The feeding of such a mobile menagerie itself presented formidable problems. Under such conditions the enemy, usually uncluttered with the paraphernalia of civilised armies, always enjoyed superior mobility. Another aspect of the problem was that reinforcements or relief could never be expected to arrive quickly.

But finally, having overcome its multitudinous problems, the Victorian army came face to face with its enemies, and gave battle. The enemies were as disparate as they were numerous, and their tactics varied accordingly. The Mutiny, the Sikh Wars and the Egyptian War of 1882 brought forth enemies trained and equipped on European lines and fighting as regulars, complete with artillery. The Zulus, formidable alike for their numbers, their mobility and their extraordinary discipline, understood only the offensive, hand to hand. At the other end of the scale, the Boers, relying on their horses and their superb marksmanship, preferred the defensive, and thought discretion much the better part of valour. The Maoris too were stout defensive fighters, using a plentiful supply of firearms from behind their stockades. In the Sudan, the Dervishes, while prefering the mass assault, showed themselves able to fight on the defensive and capable of conducting a siege. The Afghans and the tribes of the North West Frontier, adept at sniping and ambushes, were natural guerillas.

Faced with so bewildering a variety of tactical problems, complicated as they were by natural ones, it is unremarkable that the majority of Victorian soldiers were conservative in their outlook. They stuck to the methods which long experience of small colonial wars had proved successful; they had little need of innovation. In retrospect the victories of Queen Victoria's armies may seem cheap and easily bought: their very profusion lends them a sense of inevitability. The dividing line between victory and defeat was in reality narrower than statistics seem to indicate. In a remote and inhospitable land, beyond reach of rapid assistance, and amongst a savage enemy, an error of judgement on the part of any officer, any infringement of the well-tried rules, could bring about disaster, as such names as Isandhlwana or Maiwand bear testimony.

Fortunately for us, the Victorians tackled the problems of photographing their army and its campaigns with the same energy that they set about the fighting itself. The earliest photographs of a British war were taken by Surgeon John MacCosh of the East India Company's Bengal Establishment, who took a number of portraits and views during the Second Sikh War of 1848-9 and the Second Burma War of 1852-3. MacCosh used the calotype process patented by W. H. Fox Talbot in 1841, in which paper sensitized with silver salts was used to produce a negative from which a positive print could be made.

Calotype was replaced by the processes in which glass was utilized to form a negative. Felice Beato, the photographer of the Indian Mutiny and the Third China War, used the albumen-on-glass method, in which silver salts were held on the glass plate with a coating made from egg-white. The disadvantage of this method was that it required an exposure of between five and fifteen minutes, and was therefore largely confined to landscapes and buildings.

During the Crimean War the albumen process, in this case used by James Robertson, overlapped with the other glass negative method, the wet-plate or wet-collodion method, introduced in 1851. The sensitizing chemicals were held on to the plate by collodion, and while this method was superior to the albumen process in that it required an exposure of only ten seconds or less, its great drawback lay in the fact that the plates had to be exposed and developed while still wet, which meant that the photographer had to carry a complete darkroom on his travels. When Roger Fenton, who used wet-collodion, arrived in the Crimea he had thirty-six cases of photographic equipment, and his wagon was large enough to be fired on by the Russians who took it to be carrying ammunition. Wet-plate collodion endured till about 1880, and the pictures taken by the Royal Engineers in Abyssinia, and by J. Burke during the Second Afghan War, were made by this process.

Methods such as these, requiring as they did considerable technical skill and large quantities of equipment (and complicated by the problems of war conditions) put photography beyond the means of the average amateur, although in 1873 the taking of pictures was made somewhat simpler by the introduction of the dry-plate or gelatin emulsion process, which made the job of the military photographer in particular easier by virtue of the fact that the plates would keep for long periods.

The advent of simple and popular photography came in 1888, when the box camera and celluloid roll film became available. Now every officer who cared to could take snap-shots in the field, and the campaigns of the nineties were recorded in this way. Yet this too had a disadvantage, for in the hands of the amateur the box camera led inevitably to an emphasis on quantity rather than quality. Photo-albums abounded, but their contents seldom rivalled the technical and artistic achievements of Beato, Fenton, Robertson and Burke, the masters of Victorian military photography.

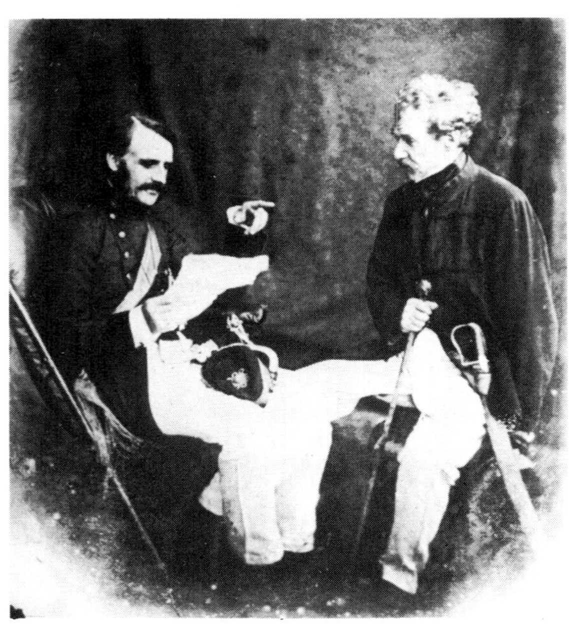

1. *Colin Campbell (1792–1863)*
Campbell was a veteran of the Peninsular War, where he served under both Moore and Wellington, and of the Walcheren expedition. He survived that disaster to soldier in Nova Scotia, Gibraltar, the West Indies and in the China War of 1842–6. During the Crimean War he commanded the Highland Brigade at the Alma and the Thin Red Line at Balaclava. He served with distinction in the Indian Mutiny, when he raised the siege of Lucknow. Between 1857–60 he was commander-in-chief India.

Campbell was the son of a Glasgow carpenter, and in the Victorian period it was rare indeed for such a person to rise to the highest ranks of the army, but in 1858 he was created Baron Clyde and in 1862, field marshal. In this photograph by Felice Beato, taken during the Mutiny, Campbell, on the right, is seen with his chief of staff, Maj Gen W. Mansfield.

2. *Robert Cornelius Napier (1810–90)*
Napier received his commission in the Bengal Engineers in 1826, and in the early years of his career was responsible for many military and civil engineering projects including road and irrigation works. He distinguished himself in both the Sikh Wars of the 1840s and again during the Mutiny, when he commanded a brigade at the relief of Lucknow. He commanded the 2nd Division during the China War of 1860 and became commander-in-chief Bombay in 1865.

His most important campaign was the Abyssinian expedition of 1867, for which he was created Baron Napier of Magdala in the following year. In 1870 he was commander-in-chief India, and became a field marshal in 1883.

3. *The Duke of Cambridge (1819–1904)*
'Royal George', cousin to the Queen, came from Hanover in 1830 and settled in England on Victoria's accession. In 1845 he became a major general, having commanded the 17th Lancers during disturbances at Leeds in 1842 and the troops on Corfu 1843–5. From this post he moved to the Dublin District and then to the command of a division in the Crimea where he was present at the Alma and Inkerman.

In 1856, on the basis of this slender military experience, he became commander-in-chief of the British army, in succession to Lord Hardinge, largely through the queen's influence. He held the post until 1895, when he was forced to resign as a result of a dispute with the secretary of state for war. As commander-in-chief he was anathema to Wolseley and all those in favour of army reform and his immensely long reign at Horse Guards was characterised by a stubborn resistance to innovation of any kind.

4. *Frederick Sleigh Roberts (1832–1914)*
Born in India, Roberts joined the Bengal Artillery in 1851 and won the Victoria Cross during the Mutiny. Thereafter he rose in the quartermaster general's department, till in 1868 he accompanied Napier's expedition to Abyssinia as AQMG. As QMG India he was an ardent advocate of the forward policy on the North West Frontier and commanded the Punjab Frontier Force in 1878. Fame came to Roberts in 1878–80, during the Second Afghan War, especially for the celebrated march from Kabul to Kandahar. But his greatest memorial perhaps, was in the affection of his soldiers, to whom he was simply 'Bobs'.

Between 1885 and 1893 Roberts was commander-in-chief India. In 1895 he became a field marshal and commander-in-chief Ireland, a post he held till he took the supreme command in South Africa, where he reversed the sorry fortunes of 1899 and forced Cronje to surrender. From 1900 till 1905 he was the last commander-in-chief of the British army.

5. *Garnet Joseph Wolseley* (*1833–1913*)
Wolseley was the most outstanding of Victoria's soldiers and was sometimes referred to as 'our only general'. He joined the army in 1852 and served with distinction in Burma, the Crimea, China, the Indian Mutiny, Canada, Ashanti, Natal, Zululand, the Transvaal, Egypt and the Sudan.

A man of immense ambition, Wolseley had greater intellectual powers than was usual in Victorian generals. He was an ardent believer in army reform and a staunch supporter of Cardwell's policies for accomplishing this. He was also an elitist and when in Ashanti in 1873–4, formed the notorious 'Wolseley Ring'—a small group of very able officers loyal to himself and his ideas whom he thereafter placed in positions of influence whenever possible. Such methods brought him into frequent disagreement with the Duke of Cambridge, whom he succeeded as commander-in-chief, 1895–9.

6. *Charles George Gordon* (*1833–85*)
Gordon, half soldier, half mystic, entered the Royal Engineers in 1852 and saw service in the Crimea. In 1860 he participated in the Third China War and stayed in the country to command the Imperial Chinese 'Ever Victorious Army' in suppressing the Taiping Rebellion of 1863–4, which earned him the sobriquet 'Chinese Gordon'. He is shown here in Chinese costume.

Between 1874–6 he ruled Egypt's provinces in Equatoria, where he traced the Victoria Nile to Lake Albert and strove, unsuccessfully, to suppress the slave trade. He returned to the Sudan as governor general in 1877 and from 1880 held a bewildering variety of posts in India, Mauritius, South Africa and Palestine.

Then in 1884, at the request of the Gladstone government, who thought that his reputation as an administrator of primitive peoples would save them the cost of a full-scale expedition, Gordon went to the Sudan once more, to extricate the Egyptian garrisons beleaguered by the Mahdists. Once at Khartoum, however, he decided that evacuation was impossible, and refusing to come away himself, defended the city for 317 days until he met his martyr's death.

7. *Redvers Henry Buller* (*1839–1908*)
Buller was commissioned in 1858 and after service in India and China went to Canada, where he took part in the Red River expedition of 1870. Here he attracted the attention of Wolseley, who made him chief intelligence officer in Ashanti. Buller won the Victoria Cross at Hlobane during the Zulu War, and was again in charge of intelligence during Wolseley's 1882 Egyptian campaign. During the Eastern Sudan Campaign of 1884 he commanded an infantry brigade at El Teb and Tamai and was chief of staff to Wolseley during the expedition to rescue Gordon. There followed a quieter period at the War Office, in Ireland and at Aldershot.

Buller enjoyed an immense reputation as a fighting general, but lacked the qualities required for high command. He was sent to South Africa in 1899 to retrieve the situation there amid great public acclaim but against his own better judgement. He suffered defeats at Colenso, Spion Kop and Vaalkrantz before relieving Ladysmith and was replaced by Lord Roberts.

His career ended tragically in 1901 when he made an injudicious reply to an attack made on him and was removed from the Aldershot command.

8. *Horatio Herbert Kitchener* (*1850–1916*)
Kitchener was commissioned into the Royal Engineers in 1871 and his early service saw him exploring and surveying in Palestine and Cyprus. In 1882 he was in Egypt, and it was in Egypt and the Sudan that his reputation was made. He accompanied Wolseley's expedition to relieve Gordon in 1884–5 as an intelligence officer, and in 1886 was made governor general of the Eastern Sudan. Two years later he was adjutant general of the re-built Egyptian army and became its sirdar in 1892.

There followed years of careful preparation, during which the Egyptian army was tempered into an instrument with which to re-conquer the Sudan from the Dervishes. Then, from 1896 till 1898 Kitchener pressed the campaign of re-conquest with a precise and ruthless efficiency, to its climax at Omdurman.

During the Boer War he was first chief of staff to Roberts, then commander-in-chief 1900–2. From then until 1909 he was commander-in-chief India.

9. *Robert Stephenson Smyth Baden-Powell (1857–1941)*
Baden-Powell was gazetted to the 13th Hussars in India in 1876 and later saw service in Zululand, Ashanti and Matabeleland. His particular interest lay in scouting and intelligence work: he was intelligence officer for the Mediterranean 1891–3 and published his *Aids to Scouting* in 1899. At the outbreak of the Boer War he went to South Africa to raise additional cavalry regiments and became a national hero commanding the garrison of Mafeking during the greatly exaggerated hardships of its long siege. This photograph was taken at Mafeking. After the war Baden-Powell raised the South African Constabulary before retiring from the army to devote his attention to the Boy Scout movement. He was inspector general of cavalry, 1903–7.

10. A mortar battery in front of Picquet House during the siege of Sebastopol, in the Crimean War. The centre gun is being loaded. Being very immobile the chief use of these medieval looking weapons was in siege operations. The barrel was normally secured to its bed at a fixed angle of 45°, the range being adjusted by varying the charge.

11. Heavy siege guns at Addiscombe, 1860. These massive muzzle-loaders, the biggest of which weighed twenty-three tons, are mounted on garrison sliding carriages. The recoil was controlled by the gun's own weight moving up the inclined slide. A point of interest is the two rows of studs which can be seen on the shells, designed to engage the rifling and rotate the projectile.

12. A mountain battery, about 1880, probably on the North West Frontier. First raised in 1851, mountain batteries proved invaluable not only on the Frontier, but in any mountainous terrain. Each gun was carried by six mules: two for the barrel, which unscrewed into breech and muzzle sections; one for the trail; another for the wheels, and two for the ammunition boxes and axle. Here the guns are ready for action, with the ammunition mules behind, the rest having been sent to the rear.

13. A 12-pounder rifled breech loader, 1896. This weapon was used by both field and horse batteries. It weighed about 6cwt, and with a service charge of 4lb, could fire a 3in shell nearly 6,000 yards. The gun crew are in field service uniform.

14. An Armstrong breech loading gun, with limber.

15. The Gatling was the first machine gun adopted by the British army, and was used in Ashanti in 1874, in Zululand in 1879—one such is shown in the picture—and later in the Sudan. The cartridges were gravity fed from the drum-like magazine to the barrels as these moved round the axis by the turning of the crank, each being fired as it reached the lowest position. The Gatling was not a popular weapon. It was clumsy and though capable of firing 300 rounds a minute, the .45 rolled brass Boxer cartridges frequently jammed.

16 & 17. In 1889 the army introduced the Maxim gun, its first recoil-operated weapon. It not only had twice the fire-power of the Gatling, but was infinitely more portable. Picture 16 shows a Maxim detachment of the 1st King's Royal Rifles on the Chitral Relief Expedition of 1895, with the gun on a stand. In photograph 17 (*facing page, above*), taken at home a year later, the 3rd Battalion's Maxims are mounted on carriages, like small field-guns. The men are wearing rifle-green uniforms with the sealskin rifle busby, in contrast with the active service khaki of the first picture.

18. A Colt gun in action at the battle of Pieters, Boer War. This was a belt fed, gas operated semi-automatic weapon, of .303 in calibre. It is seen here mounted on Lord Dundonald's galloping carriage, which was drawn by a single horse, without a limber.

19. During the Boer War the need for greater long-range fire-power led to the adoption of some naval ordnance. Here a 6 in gun is mounted on a railway truck to give it greater mobility. The photograph was taken in March 1900 at Modder River.

20. A pom-pom and limber, South African War. Designed by Sir Hiram Maxim, the pom-pom was a recoil operated automatic gun firing 1lb explosive shells. In the late 1890s the British Government turned the weapon down and the manufacturers sold it to the Government of the Transvaal instead. When the Boer War started the British army found itself on the wrong end of these formidable weapons and was hastily equipped with them.

21 & 22. These two pictures typify the extremes of Victorian military thinking. In the one, cavalry swords are being sharpened in readiness to deal with the Boers: in the other, taken during the Tirah Campaign of 1897, a primitive type of rocket is shown in use. This is probably a Hale's rocket, introduced in 1867. They proved to be ineffective, and more dangerous to the users than the enemy and were not employed again after the Tirah expedition.

23. Sir George Brown and his staff in the Crimea. Their uniforms, with frock coats and cocked hats, would have been at once recognizable to the armies of the Napoleonic era.

24. Infantry in the Crimea, showing a variety of uniforms. The soldier on the left is in great-coat and 'pork-pie' forage cap. The three central figures are wearing Albert shakos and coatees. The two seated men to the right of the picture are in forage caps and shell jackets.

25. A cornet of the 11th Hussars, photographed in the Crimea. The tunic was blue and laden with gold lace. The crimson overalls too, had a broad gold stripe. The brown busby, with crimson bag, was surmounted by a huge white plume.

26. The uniform of this trooper of the 5th Dragoon Guards was somewhat simpler. The tunic was scarlet with green facings, the trousers dark blue with yellow stripe. The helmet was brass and the sheepskin plain black.

27. A group of the 101st Royal Bengal Fusiliers (later the 1st Royal Munster Fusiliers) 1863–4. The regimental number and the fusilier grenade badge can be seen on the forage caps. The men wear single-breasted tunics with slashed cuff-panels. The pouch on the white buff cross belt held percussion caps for the Enfield rifles with which the soldiers are armed.

28. Men of the 86th Regiment (2nd Royal Irish Rifles) at Port Elizabeth, Cape Colony, 1867. Their uniforms, with the white tunics and the white covered forage caps with the neck-flaps, are reminiscent of those worn during the Mutiny in India.

29 & 30. Fears of a French invasion in 1856 led to the establishment of numerous volunteer corps including Rifle Volunteers, Light Horse, and Mounted Rifle Volunteers, Artillery and Engineers. Government response to this patriotic enthusiasm was to give virtually no material or financial assistance: most weapons, all clothing and equipment and in the case of mounted units, horses as well, were provided by the volunteers themselves.

Although some of the yeomanry cavalry regiments chose lancer or dragoon style uniforms, most went for a simplified hussar style, of which this (picture 29), the uniform of the Warwickshire Yeomanry Cavalry, is an example. The rifle volunteers too, displayed a variety of dress, but the Queen's Westminsters shown on the facing page (picture 30) are of the typical grey colour and pattern. Both photographs date from about 1870.

31. A group of the Royal Norfolk Regiment, taken in the mid nineties. The scarlet frock tunic, in this case with white facings, is interesting in that it has a flap on the left shoulder, to prevent rifle oil from staining the tunic itself. The trousers were of blue serge, with black leather leggings. The helmet was of the blue cloth covered cork type introduced in 1878. The men are wearing the Slade Wallace 1888 pattern valise equipment in white buff leather.

Facing page: (*above*)
32. Sergeants of the 1st Royal Norfolk Regiment, some wearing glengarries and some in forage caps on which can be seen the regiment's Britannia badge. The frock is of an experimental pattern introduced in 1884 when attempts were being made to find a suitable pattern khaki uniform for use in the United Kingdom. This particular style was soon discontinued.

(*below*)
33. A group of the 1st Battalion Seaforth Highlanders, photographed during manoeuvres in the New Forest in 1895. They are wearing the Mackenzie tartan and their doublets have the pocket-like flaps known as Inverness skirts. The bugler, second from left, is distinguished by his wing-epauletts. The two recumbent men are in undress, wearing glengarries.

34. This photograph, taken in 1896, shows the variety of uniforms within a single Scottish regiment —the 1st Royal Scots Fusiliers. To the left is the drum-major, his shoulder belt embroidered with the regimental battle honours. Next to him is a private in full-dress, wearing the black sealskin fusilier cap. The RSM, in the centre, is in undress, with forage cap and cut-away frock tunic. The piper is in full-dress. To the right stands a colour sergeant, his doublet clearly showing the Inverness skirts.

35. A private of the Seaforth Highlanders ready for kit inspection, 1896.

36. A trumpeter of the 13th Hussars at Aldershot, before going out to South Africa, 1899. Gone are the sartorial splendours of earlier years and in their place khaki, spare boots, forage-bag and picketing gear. An interesting point is that the trooper carries both trumpet and bugle. The bugle was used to sound calls when mounted and in the field; dismounted calls were sounded on the trumpet.

37. The drum horse of the 3rd King's Own Hussars, 1895. The kettle drummer wears the uniform of a hussar sergeant, with the interesting addition of an engraved silver collar, presented in 1772 to the 3rd Dragoons (from which regiment the 3rd Hussars descended) by the wife of the colonel. The drums were captured by the 3rd Dragoons at Dettingen, in 1743.

38. A staff corporal farrier of the 1st Life Guards, 1896. The farriers of all regiments wore blue and in the case of the Life Guards were further distinguished by the black plume and the absence of a cuirasse. The farrier's axe is carried at the 'Advance'. At this period its function was purely ceremonial, but its original purpose was to despatch horses too badly wounded to move.

39. A captain of the 17th Lancers, in review order, 1896. The breeches and tunic were dark blue, with white plastron and schapka plume. The skull and crossbones badge is clearly visible on the shabraque and attached to the right stirrup-iron can be seen the leather boot for resting a lance-butt.

Facing page
40. Officers of the 9th Lancers, 1894, in full-dress. The tunic was blue, the gold stripes being replaced in 1894 by yellow cloth. The plume feathers were black and white.

41. Officers of the 12th Lancers in a variety of undress uniforms including tunics and stable-jackets, overalls and breeches, pill-box and field-service caps, 1884.

42. A battery sergeant major, Royal Horse Artillery, 1896. His busby was of black sealskin, with red bag, white horse-hair plume and yellow cap-lines. The blue shell jacket, with red collar, was decorated with gold cord loops; the breeches, also blue, had a broad red stripe. The boots were of the Hessian style.

43. Colour sergeants of the Rifle Brigade, 1898, dressed for campaigning in the Sudan in uniforms shorn of all peacetime elegance. The white cork helmet was worn with khaki cover and curtain. The tunic, quite plain except for two breast pockets, the trousers and puttees were all khaki. The Slade Wallace equipment was in brown leather.

Facing page
44. Irish Guardsmen, 1901, in marching order. By this time the black bearskins, scarlet tunics, blue trousers and buff-white valise equipment were for peacetime conditions only. The shamrock can be seen on the collars.

TRAINING

45. The 35th (Royal Sussex) Regiment on parade at Mooltan in 1866. Many of the men are wearing their Crimean and Indian Mutiny medals.

46. A group of 12th Lancers on the parade ground at Bangalore, 1885.

47. At Port Elizabeth in Cape Colony, 1871, troops prepare to leave camp on a route march. This type of training was of critical importance in an age when infantrymen moved everywhere on foot.

48. Digging shelter trenches, 1887. The army at this time seldom had occasion to use trenches except in siege operations, such as in the Crimea, or during the Mutiny.

49. A British Square. In this picture, taken by J. Burke at Peshawar in 1885, is shown the 2nd Battalion Dorsetshire Regiment. The front rank kneels, the second rank stands, both with fixed bayonets. Behind are the company officers and NCOs, with mounted officers and band in the centre. In action a square would also shelter medical staff, the ammunition reserve, baggage and transport.

50. Cavalry were required to be proficient with their carbines as well as with the *arme blanche*. One half of a lancer troop are practising at the butts while the others hold the horses and lances. Southern India, 1885.

51. Horses were sometimes trained to lie down to give cover to their riders when firing dismounted. In this picture, taken in 1896, a corporal of the 17th Lancers is taking part in a training exercise. His weapon is a Martini-Henry carbine.

52. The lance was a difficult weapon to manage and skill in its use demanded thorough training. Here men of the 12th Lancers are engaged in dismounted lance drill at Ballincollig, Ireland, about 1895.

53. Three troopers of the 13th Hussars on exercises at Aldershot, prior to going to South Africa in 1899. They are in service order and carry Lee Enfield cavalry carbines.

54. Artillery Practice, 1866.

55. Royal Engineers laying a bridge, 1870. This wood and rope structure was capable of supporting a gun and limber.

56. 3rd Grenadier Guards at volley-firing practice, in about 1890.

57. Bayonet drill, around 1890.

58. A fencing class, about 1890. In cavalry regiments swordsmanship was still an important skill.

TRANSPORT AND COMMUNICATIONS

59. The Victorian army was animal powered, and the horse ubiquitous. Here a field artillery gun team wait to go into action, 1895.

60. A sergeant of the Gordon Highlanders Mounted Infantry, 1896. The need for greater infantry mobility became evident in the late seventies, and in 1884 training schools were set up for mounted infantry at Aldershot and the Curragh, home establishment regiments sending an officer and thirty-two men each. Mounted infantry were not cavalry, for they used their horses only for transport, and dismounted to fight. Standards were high and sections were able to gallop 1000 yd, dismount and run 100 yd and fire three accurate rounds, in six minutes. In 1896 a Highland Company of Mounted Infantry was formed, composed of detachments from the Black Watch, Seaforths, Argylls and Gordons.

61. The maintenance of saddlery was a matter of the greatest importance to both the men and horses of a cavalry regiment. While these saddlers of the 17th Lancers stitch away, a sergeant directs work on an officer's shabraque, 1896.

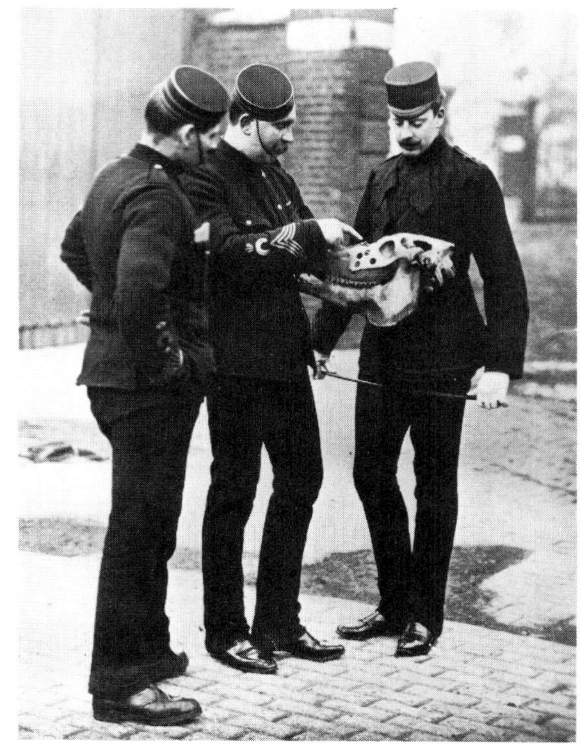

62. Care of the army's vast number of horses was of vital importance, and absorbed the attention of many personnel. In 1881 the Army Veterinary Department came into being to supervise this important work. Here two farriers of the 1st Life Guards discuss a point of equine anatomy with an officer, 1896.

63. Shoeing-smiths at work on the veldt during the Boer War, about 1901.

64. A wagon of the Military Train, 1866. The Military Train was formed in 1856 from the Land Transport Corps which had been raised the year before for service in the Crimea. In 1857 its strength was increased from three to six battalions, each of four troops. Battalions of the train served with distinction in the Mutiny, the Third China War and in New Zealand and when not actually engaged in transport duties were employed as cavalry. In 1869 the Control Department took over the duties of the Military Train and the Commissariat and absorbed the train's officers. The other ranks went into the first Army Service Corps.

65. An ammunition mule of the 2nd East Lancashire Regiment, 1896. An infantryman on active service was allocated 185 rounds of rifle ammunition. 100 rounds he carried himself, 65 rounds were carried by the mules as battalion reserve and a further 20 rounds in the battalion baggage wagons. Two mules were allotted to a company, each carrying two ammunition boxes containing 1100 rounds apiece, and together weighing more than 150lb.

66. No 10 Battery Mountain Artillery advancing in open order near Pietermaritzburg, Natal. The battery's 7-pounder guns and ammunition were carried on mules. The photograph shows vividly what a late Victorian army looked like on the move. At the time this photo was taken, the battery had one section on active service in the Matabele campaign of 1896.

67. Oxen were widely used as draught animals in southern Africa, though their surprising delicacy made for slow progress: they died if overworked or if grazing was inadequate. Getting a wagon over a difficult obstacle often required it to be double spanned; here no less than fourteen span of oxen are being used to pull a wagon through a drift. Boer War.

68. Oxen were also used in India. Here they are drawing artillery and wagons, 1899.

69. Elephants were employed in India to pull heavy guns. There were four such batteries in India; this photograph was taken at Secunderabad in 1896. Each battery was commanded and manned by Royal Artillery personnel and comprised six guns—four of the 40-pounder muzzle-loaders shown here, and two 6.3 in howitzers. Transport for each battery was on an appropriately large scale: 262 bullocks were required to haul the equipment and ammunition wagons, attended by numerous Indians.

70. In India the wounded were often transported in a type of litter known as a doolie. The occupant of this one is having a morphia injection, Tirah, 1887.

71. Towards the end of the century infantry battalions experimented with cyclist sections as a possible means of improving mobility, or as a contemporary journal put it, 'for any employment in which speed, celerity and general smartness are requisite'. These cyclists were on Lord Methuen's staff in South Africa.

72. HMS *Himalaya* in the Suez Canal, bringing the 39th Foot home from Alexandria, August 1886.

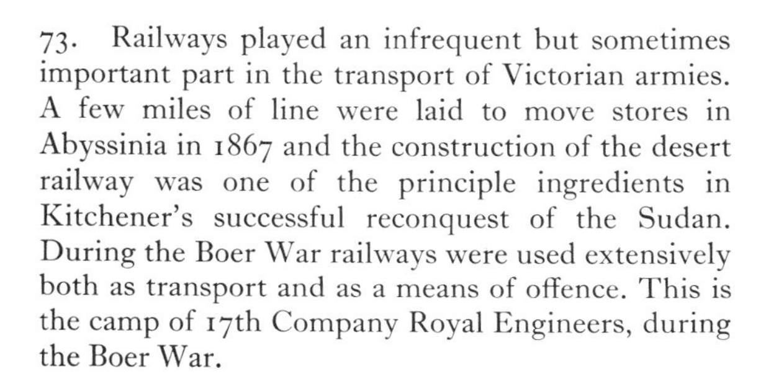

73. Railways played an infrequent but sometimes important part in the transport of Victorian armies. A few miles of line were laid to move stores in Abyssinia in 1867 and the construction of the desert railway was one of the principle ingredients in Kitchener's successful reconquest of the Sudan. During the Boer War railways were used extensively both as transport and as a means of offence. This is the camp of 17th Company Royal Engineers, during the Boer War.

74. The army began experimenting with balloons in 1878, at Woolwich Arsenal. They first saw field service in Bechuanaland in 1884 and again the following year in the Eastern Sudan. In 1890 the Royal Engineers formed a Balloon Section at Farnborough. Four sections were employed during the Boer War. The one shown here was used for observation by the defenders of Ladysmith, 1899.

75. The electric telegraph, invented in 1837, was first used by the British army in the Crimea. The first Royal Engineers units to specialize in signalling were created in 1870. Here we see a telegraph section photographed during the Second Afghan War of 1879–1881.

76. The heliograph was a less sophisticated signalling system and dependent on good visibility. This helio section is relaying news of the battle of Dargai to Fort Lockhart. Tirah, 1897.

77. Semaphore was still in use during the Boer War.

78. War correspondents—Mr Hutchinson of *The Times* and Melton Prior of The *Illustrated London News*—in Tirah, 1897. Following W. H. Russell's reporting of the Crimean War, correspondents covered most major expeditions, and some at least were enthusiastic participants in the fights they were to write about.

RECREATION

79. The immense variety of service in the army at this period offered unrivalled opportunities for big-game hunting and many officers spent their leaves exploring remote regions in search of trophies. This collection, shot in India between 1891 and 1896, is recorded in the original album as being part of a total bag of 3 elephant, 19 gaur, 9 tiger, 6 bear, 6 ibex, 26 deer and antelope, a jungle sheep, plus pig and innumerable species of bird.

80. All forms of equestrian sport were naturally popular and there were few places in which the army failed to organize a local race meeting or gymkhana. In India pig-sticking was popular and was run on a competitive basis for handsome prizes.

81. The 12th Lancers' team, Bangalore polo tournament, 1895.

82. The object of tent-pegging was to spear and remove a large wooden peg driven securely into the ground. The two horsemen on the left have missed their pegs, but the rider to the right has successfully impaled his on his lance point. The photograph shows an officer and men of the 2nd Mounted Infantry at Harrismith, South Africa, after the end of the Boer War.

83. Peshawar Vale Hunt, in about 1890. At Poona, Peshawar, Lahore and Quetta were regular hunts whose servants wore scarlet and whose hounds were sent out from England. Elsewhere there were 'bobbery' packs, collections of any available dogs whose hunting prowess was indifferent, but which gave good sport. In India the jackal rather than the fox was the quarry.

84. The cricket team of the 12th Lancers, India, in about 1885.

85. A mixed badminton party, India, in about 1864.

86. A bicycle race at the Karachi gymkhana, 1898.

87. While service in India had many advantages for officers with sporting propensities, it had few for their wives, for whom the long boredom of the months was punctuated only by the annual trek to the hills for the hot weather. Amateur theatricals were a popular form of self-entertainment: this production of *Court Cards* took place at Ootacamund in 1895.

88. A luncheon party at the Camp of Exercise, Delhi, 1886.

89. A regimental band giving an open-air recital at Hampton Court, 1864.

90. Barrack room law, 1896. A trooper of the Blues before a barrack room court-martial for disturbing his sleeping comrades late at night. The horse-play disguised a real penalty. Though he escape the axe, a blanket-tossing or a ducking in the horse-trough might await, or in this case, the bill for the whisky cask, adding a damaged pocket to a wounded pride.

91. Mop-fighting, part of the Imperial Yeomanry sports during the Boer War.

92. The 2nd Royal Sussex Regiment tug-of-war team, Ferozepore, 1892.

93. 3rd Rifle Brigade boxing competition, Kuldanna, India, 1896.

94. Highlanders of Wolseley's army sightseeing after the battle of Tel-el-Kebir, during the Egyptian campaign of 1882.

CAMPAIGNS

THE CRIMEAN WAR (1854–5)

Russia sought a foothold in the eastern Mediterranean at the expense of the moribund Turkish empire and it was British and French policy to prevent such Russian expansion by supporting Turkey. Tsar Nicholas I, who demanded protective rights over Orthodox Christians living in Turkish territory, invaded Moldavia in support of his claim and Britain and France declared war in support of their ally.

In September 1854, British and French armies landed in the Crimea with the intention of seizing the Russian naval base of Sevastopol. They defeated the Russians at the Alma, failed to capture Sevastopol quickly, and so settled down to besiege the town, the British taking Balaclava as the port through which to draw their supplies.

On 25 October a Russian field army attempted to attack Balaclava and there followed one of the most famous battles in British history, with the charges of the Light Brigade and the Heavy Brigade and the 'Thin Red Line'. Another Russian attack was repulsed at Inkerman on 5 November.

The hardships endured in the long seige that followed were terrible on both sides and the Russians eventually abandoned Sevastopol to the Allies. Peace was signed in March, 1856.

95. The harbour at Balaclava. In the crowded and unsuitable anchorage large quantities of desperately needed supplies were left to deteriorate through mismanagement and the difficulties of getting them up to the camp before Sevastopol.

96. The appalling hardships endured by the troops during the Crimean winter are one of the most notorious aspects of the war and caused outrage when the facts became known at home. This picture by the famous Crimean photographer Roger Fenton shows men of the 68th in winter clothing.

97. Another Fenton picture, showing the 57th Foot drawn up in front of the camp.

98. Survivors of the Charge of the Light Brigade—the remnant of the 13th Light Dragoons the morning after the charge.

99. Inside the Redan, one of the key points in the defences of Sevastapol. The photograph is by Robertson.

British expansion in India in the 1840s and early fifties was rapid and accompanied by the introduction of British law and the abolition of ancient Hindu customs such as *sutee*. There was widespread unrest amongst the Indian civil population and it was shared by the sepoys of the East India Company's Bengal Army, who greatly outnumbered the white troops.

This smouldering disaffection was fanned into active hostility by the celebrated incident of the greased Enfield rifle cartridges, which was regarded by both Muslim and Hindu sepoys as an attempt to pollute them, a first step to their forcible conversion to Christianity. In May 1857 sepoys at Meerut refused to handle the cartridges and were court-martialled. Their comrades mutinied, murdered their British officers and marched on Delhi, where an ancient relic of the Moghul dynasty, Bahadur Shah, was proclaimed emperor.

In June a British force arrived before Delhi and commenced a siege. In September, in the face of great difficulties and heavy odds, the city was retaken by storm. At Cawnpore, the garrison held out till June, when, greatly reduced in numbers they surrendered on a treacherous promise of a safe conduct. The mutineers massacred the survivors, including the women and children, which brought a note of bitterness and savagery into the fighting from that time on.

In Oudh the British fell back on Lucknow, which was relieved in September by Havelock and Outram. They were promptly invested along with the garrison, till the second relief of the Residency by Sir Colin Campbell in November, after which a further six months were required to eliminate the remaining mutineers.

100. Part of the barracks held by the Cawnpore garrison. The Residency at Cawnpore was surrounded by the mutineers led by Nana Sahib on June 6th, 1857. The 800 defenders, under Maj Gen H. Wheeler, held out till their numbers were reduced to some 250, when they surrendered under promise of a safe-conduct to Allalabad. Nana Sahib, however, intended treachery, and a brutal massacre of the English women and children followed.

101. The gateway and banqueting room of the Lucknow Residency, which was used during the siege as a hospital.

102. A group of 1st Madras Fusiliers (later the 1st Royal Dublin Fusiliers) photographed at Lucknow after the second relief.

103 & 104. The massacre at Cawnpore provoked both at home and in India quite un-British feelings of hatred and ferocity and thereafter mutineers received no quarter. These photographs, like the others shown here of the Mutiny, were taken by Felice Beato and show some results of this spirit of revenge. Top: Mutineers being hanged. Bottom: The interior of the Secundra Bagh, a fortified palace in Lucknow. It was stormed on 16 November 1857 by Sir Colin Campbell's relief force, which lost heavily in the process. The building was held by some 2,000 mutineers who were wiped out by the 93rd Highlanders and whose remains are scattered in the courtyard.

British relations with China were strained from the beginning of Victoria's reign, as a result of British unscrupulousness over the opium trade on the one hand, and the Chinese refusal to admit British diplomats to Pekin on the other. In 1856 in a wave of xenophobic excitement the Chinese murdered a French missionary, seized the crew of a British registered vessel on charges of piracy and closed Canton to foreign trade.

Lord Elgin was sent to China with a force of 1,500 men under General Ashburnham and in December 1857, took Canton. In April the following year Elgin and Ashburnham captured the Taku forts at the mouth of the Peiho river and forced the Chinese to sign the Treaty of Tientsin, which provided for the opening of more ports to foreign trade, the protection of missionaries and the right of foreign diplomatic missions to reside at Pekin.

In June 1859, however, British and French diplomatic staff were refused permission to travel to Pekin and a naval force under Admiral James Hope was repulsed at the Taku forts. In April 1860 a joint British and French force was despatched to China, the British troops commanded by Lt Gen Sir Hope Grant. In August the Taku forts were taken by storm and the allies advanced via Tientsin on Pekin, fighting battles at Chau-chia-wan and Pal-le-chiao on the way.

The war was brought to a satisfactory conclusion, from the British viewpoint, and the Chinese were obliged to ratify the Tientsin Treaty under the title of the Treaty of Pekin.

105. A triumphal arch on the Honan Shore erected by the Chinese to celebrate their diplomatic victory over the British in 1847 and destroyed by the British in 1858. The photograph is by Felice Beato.

106 & 107. These photographs, both by Beato, show one of the Taku forts which guarded the mouth of the Peiho river and the route to Tientsin and Pekin. The defences were formidable, consisting of a moat, a ditch filled with sharpened stakes and a crenellated wall. The top picture shows the North Gate and the breach stormed by the British on 21 August 1860. The lower photo shows the interior after the assault.

108. Inside the Peh-tang fort, where the British set up an HQ. It was taken soon after the fall of the Taku forts, but without fighting.

THE ABYSSINIAN CAMPAIGN (1867–8)

In 1864, Theodore III, the paranoic king of Abyssinia, imprisoned the British Consul and his staff as the result of an unintentional slight, with some missionaries and others, a number of whom were barbarously treated. Lengthy diplomatic correspondence failing to secure the prisoners' release, stirring public opinion obliged the government to consider direct intervention.

The difficulties were immense; Abyssinia was both remote and unknown, but it was decided to send a force of 1,300 British and Indian troops from Bombay under Sir Robert Napier. From its base at Zula on the Red Sea, Napier's force advanced across 400 miles of mountainous country, entirely deficient in roads, to Theodore's stronghold at Magdala. A brisk action was fought at Arogee and the prisoners brought to safety. Magdala was then stormed, Theodore shot himself and Napier's force withdrew. As a rescue operation, the campaign was entirely successful, though very expensive.

109. Annesely Bay, on the Red Sea. Here some of the 291 transports used in the campaign are anchored, to discharge their cargoes for the base camp at Zula. The shallowness of the water necessitated the building of jetties 900 yd long. One can be seen on the left of the photo.

110. The base camp at Zula, which plague and the lack of water soon rendered insanitary and unpleasant. Here stores of all kinds were accumulated and to assist in forwarding them inland, the Royal Engineers laid some fourteen miles of 5ft 6in railway, part of which can be seen. In the right foreground are quantities of pack saddles.

111. Undul Wells, in the Sooroo Pass, on the route to Magdala. In this pass a number of followers and baggage animals were washed away by flood-water.

112. Napier and a group of Royal Engineers officers.

113. The Kafir-Bur Gate, Magdala. The troops on the Abyssinian expedition were variously dressed. These men are in the recently introduced khaki; others wore the traditional scarlet.

THE ZULU WAR (1879)

In the 1870s the Zulu people presented a spectacle unique in black Africa. Under their savage and despotic king, Cetewayo, the whole fabric of the nation had been organized around the army—more than 40,000 strong—into which the entire male population was compulsorily drafted. This great force was formidable not only for its size, but for its discipline and tactical sophistication.

To the South African High Commissioner, Sir Bartle Frere, it appeared that the Zulus might at any moment become embroiled with the Transvaal Boers and start a full-scale native war. In any case, the Zulu impis constituted a permanent threat to the colonists of Natal. After a series of frontier incidents, Frere sent Cetewayo an ultimatum demanding the disbandment of the Zulu army.

The ultimatum was ignored, and in January 1879 an army of British, colonial and native troops 12,000 strong, under the command of Lord Chelmsford, invaded Zululand in three columns. Chelmsford's own column met with disaster at Isandhlwana, the worst in Victorian military history, when some 1,300 British and native troops were massacred. This was immediately followed by the famous defence of Rorke's Drift. Of the other two columns, one was besieged and the other fought two battles, one of which, Hlobane, was another disaster.

After a period of reorganization and with strong reinforcements, Lord Chelmsford renewed his advance on Ulundi, the Zulu capital and here, in July, Zulu military power was at last broken. Cetewayo was captured some time later.

114. Isandhlwana. Here on 22 January 1879, a Zulu impi 20,000 strong fell upon the unprepared British camp and 858 British and colonial troops together with nearly 500 native levies were massacred. The scale of the disaster may be gauged from the casualty figures of the 24th Regiment. Out of the single company of the 2nd Battalion present, all five officers and 178 men fell. The principal infantry unit in the camp was the 1st Battalion of the same regiment, of which all sixteen officers and 400 out of 403 other ranks were killed. In the foreground of the photograph can be seen the wagons, the remains of slaughtered oxen and the scattered debris of the battle.

115. If Isandhlwana was the Victorian army's greatest disaster, Rorke's Drift was the scene of its most celebrated defensive action. On the afternoon of 22 January, some 4,500 warriors of the impi that had destroyed the camp at Isandhlwana attacked the mission station at Rorke's Drift, which Lord Chelmsford's column was using as a hospital and store. The garrison totalled 104 men of all ranks fit for duty, composed principally of B Company 2nd/24th, and was commanded by Lt Chard of the Royal Engineers.

The Zulus pressed their attacks with the utmost ferocity and persistence, much of the fighting at the walls being hand to hand, till the early hours of the following day. The Zulu casualties were at least 400 dead; of the British, fifteen were killed and scarcely a man unwounded. Eleven VCs were awarded for this action. This photo, taken some time after the battle, shows the burned out hospital, and to the left, the monument to the dead.

116. A group of Royal Engineers in Zululand. Lt Chard, wearing his Victoria Cross, has been arrowed on the original picture.

117. Lord Chelmsford's army bivouacing the night before the final battle of the war at Ulundi. The weather was cold and misty, and the troops, without tents or personal kit, spent an uncomfortable night.

118. The funeral parade of HRH The Prince Imperial, 2 June 1879. Chelmsford had sufficient military problems to worry him without the arrival in Zululand of Louis Napoleon, the exiled Prince Imperial of France. Although officially a spectator, Louis thirsted for action and after several rash adventures was killed while on a reconnaissance. His death brought a fresh public outcry upon the head of Lord Chelmsford, already oppressed by the Isandhlwana incident.

Fears of Russian influence in Central Asia as a threat to India, which had brought about the First Afghan War of 1838–42, increased in the 1870s and Lord Lytton, the Viceroy, was instructed to establish a British Mission in Kabul. The Amir of Afghanistan, Sher Ali, refused entry to the mission, which was interpreted as proof of Russian intrigues. In 1878 war was declared and three columns successfully invaded Afghanistan. The force under Sir Frederick Roberts won a small but brilliant action at Peiwar Kotal.

The amir fled and his successor, Yakub Khan, agreed to permit a British envoy to reside at Kabul and control Afghanistan's foreign affairs. In September 1879, however, the Envoy, Sir Louis Cavagnari and his escort were massacred. Lt Gen Sir Donald Stewart re-occupied Kandahar and Roberts fought his way through the Kurram Pass to Kabul, where he occupied the great cantonment of Sherpur. He was immediately invested and on 23 December beat off a furious attack by 100,000 Afghans. Yakub Khan abdicated and was replaced by a British protégé, Abdur Rahman.

In May 1880, Roberts was joined by Stewart, who had fought his way from Kandahar, but in July Ayub Khan, the brother of Yakub Khan, inflicted a crushing defeat on a British force at Maiwand, and laid seige to Kandahar. Roberts set out from Kabul with a relief column and in a celebrated march covered the 300 miles in twenty days. He defeated Ayub Khan and raised the siege. Peace was restored and Abdur Rahman confirmed in power. The British withdrew, but remained deeply suspicious of Russian intentions.

119. Officers of the 67th Regiment.

120. This photograph of the fort at Ali Musjid in the Khyber Pass, gives a good idea of the terrain and conditions on the North West Frontier.

121. The defences of the Sherpur cantonment at Kabul. This is one of the many fine pictures taken of the campaign by J. Burke.

122. A field hospital.

123. Another Burke picture, showing captured Afghan artillery.

The British first became involved in the tiny Himalayan country of Chitral in 1876 when the Chitralis, under pressure from more powerful neighbours, placed themselves under the protection of the Maharaja of Kashmir, himself a vassal of the Government of India.

In 1892 the death of the Mehtar of Chitral was followed by long and bloody disputes amongst his relatives for possession of the throne, which was eventually seized by one of his sons. In July 1895, Surgeon Major George Robertson, the Political Agent in Gilgit, arrived in Chitral with 400 Sikh and Kashmiri troops and half a dozen British officers to stabilize the situation. Robertson dethroned the incumbent ruler and enthroned his twelve-year-old brother Sujah-ul-Mulk.

The Mehtarship however, was coveted by Sujah-ul-Mulk's uncle, Sher Afzal, who assembled an army and prepared to contest the issue. Robertson sent the Kashmiri troops to seek Sher Afzal, but they suffered heavy casualties, including two of their British officers. Robertson's force, now numbering 343 fit men, retired into the fort at Chitral, which was closely invested.

At Peshawar Maj Gen Sir R. Low assembled a relief force of 15,000 British and Indian troops. But this large column could advance only slowly, and Low was forestalled by the heroic efforts of Lt Col J. Kelly and the 32nd Sikh Pioneers, who advanced from Gilgit through appalling Himalayan conditions and successfully raised the siege of Chitral, which had lasted forty-seven days.

124. The north-west corner of Chitral fort. The walls and towers were stoutly constructed of timber, stone and mud. The tribesmen succeeded in burning one of the towers and attempted to breach the walls by mining. They were discouraged by a sortie from the defenders, who suffered additional hazards in that the latrines and the water supply both lay outside the walls.

125. Conditions within the fort at Chitral were uncomfortable, and for the relief column hardly less so. This is an officers' mess.

126. Mountain guns in action.

127. A rope bridge at Drosh. This photo well illustrates some of the difficulties facing a Victorian commander. Moving a force across such a river, with the original bridge destroyed and an enemy entrenched on the opposite heights would be a formidable and time consuming problem.

TIRAH (1897–8)

In 1897, the year of Queen Victoria's Diamond Jubilee, a Moslem holy-man, known as the Mad Mullah to the British, fomented sufficient trouble on the North West Frontier to precipitate the Great Frontier Rising. The tribes rose virtually en masse from Swat to Baluchistan, and control of the Khyber was temporarily lost. Several forces were put into the field to quell the trouble—in Malakand, in the valley of Swat, against the Mohmands and in the Tochi valley.

These measures proved insufficient. Both the Khyber and Kohat passes lay within the territory of the Afridis, largest and most turbulent of the tribes, who, with their almost equally unruly neighbours the Orakzai, lived in the mountainous area south-west of the Peshawar valley known as Tirah. No Englishman had ever penetrated Tirah, but it was resolved to despatch a strong expedition to exact retribution from the tribesmen and to dictate peace conditions from the heart of the country.

Lt Gen Sir William Lockhart was appointed to command the force, composed of some 35,000 British and Indian troops—more than had been sent to the Crimea—and operations began in mid October, 1897. Speed was essential, for Lockhart had to perform his task and extricate his army before the onset of winter made campaigning impossible. On 20 October a ferocious though long forgotten fight took place at Dargai, in which a strong Afridi position was stormed frontally, with heavy casualties. Thereafter, as the advance continued, the tribesmen relied upon guerilla tactics.

Having reached the Maidan valley, the heart of Tirah, Lockhart's force set about chastising the tribes by the normal frontier expedient of burning houses and villages and destroying crops, fodder, grain and fruit trees. In November Lockhart began his withdrawal on Peshawar, before snow blocked the passes. As the weather deteriorated the retiring columns were harassed incessantly.

Few of the Afridi clans had submitted; hardly any of the rifles demanded as fines had been handed in. The tribes were only temporarily subdued and even the queen wondered whether it had all been worthwhile.

128. A section of the King's Own Scottish Borderers in action.

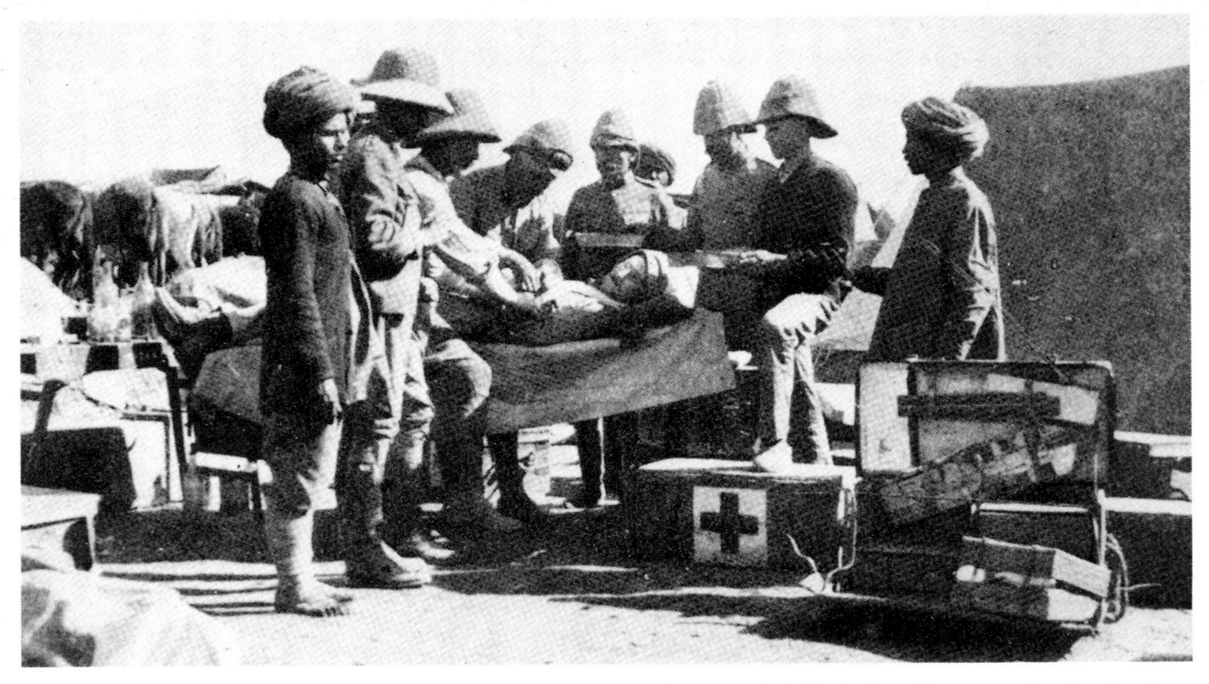

129 & 130. The cost of frontier warfare:
In a field hospital an officer of the Dorsets has an operation on a wounded arm.
Burning villages near Bagh. The razing of villages and crops was a standard procedure on frontier campaigns, though in Tirah it was carried out with more than usual thoroughness. The morality and the expediency of this policy was hotly debated in both military and civil circles. Some saw in the destruction of the tribes' slight property and few possessions the only effective way of punishing them. Others held that such methods were not only barbarous, but made the tribes more intractable. The only certain thing was that the effects of such visitations were short-lived.

131. General Lockhart, Sir Richard Udney and General Nicholson announcing peace terms to the tribal leaders.

RECONQUEST OF THE SUDAN (1896–9)

For eleven years after the death of Gordon at Khartoum in 1885 and the failure of Wolseley's expedition to save him, Madhism triumphed in the Sudan while in Britain the humiliation rankled. In 1896, Lord Salisbury's government, prompted by public feeling, alarmed at French encroachment on the upper Nile, and aware that the Egyptian economy could now finance a war, resolved upon the reconquest of the Sudan.

The Sirdar of the Egyptian Army, Maj Gen Sir H. Kitchener, was placed at the head of a mixed British, Egyptian and Sudanese force which grew steadily as operations progressed to a total of nearly 26,000 men.

Omdurman, the centre of Dervish power, lay more than 1,000 miles from Cairo, and the difficulties were formidable. Along the twin arteries of the Nile and the railway, built over the Nubian desert across the great loop of the river, the army methodically advanced on points further and further south. Progress was slow, governed by the annual rise and fall of the Nile and the speed at which the railway could be extended.

The first general action was fought at Firket in June 1896. Abu Hamed fell the following year. In April 1898 the battle of Atbara took place. Kitchener's army gradually closed upon Omdurman, where the decisive battle of the war was fought in September, 1898. It was not until the end of 1899, however, that Dervish resistance finally collapsed when Sir Reginald Wingate caught and killed the Kalifa in a last skirmish in Kordofan.

132. Kitchener made extensive use of gunboats and armed steamers and a great deal of labour was expended in getting them up the Nile cataracts. They were invaluable not only on transport and communications duties, but added greatly to the army's fire power, taking part in numerous actions, including the battle of Omdurman. Here the gunboat *Thafir* is moored at the Atbara camp.

133. A locomotive of the desert railway takes on water, Sudan, 1898. The pushing of the railway across the Nubian desert was certainly the greatest engineering feat undertaken by the army in the nineteenth century and was a vital factor in Kitchener's successful reconquest of the Sudan.

134. Although the desert railway greatly facilitated the movement of troops, supplies and equipment, the Nile played a dominant role throughout the war. Here, a 40-pounder is man-handled aboard a native sailing craft.

135. Infantry and machine guns in action at the battle of Omdurman.

136. After Omdurman the army pressed on into Khartoum. Here troops inspect the remains of Gordon's palace.

THE BOER WAR (1899–1902)

The long-standing grievances between the Boers and the British in South Africa, which had already given rise to the Transvaal War of 1880–81, again erupted into war in October 1899 when the Boers invaded Natal and laid siege to Ladysmith, Kimberley and Mafeking. Within a month Redvers Buller arrived in South Africa with reinforcements and set out to relieve Ladysmith. Expectations that the Boers would soon be crushed were shattered in December and in January 1900, when the irregular South African farmers with their highly accurate marksmanship inflicted the humiliating disasters of Stormberg, Magersfontein, Colenso and Spion Kop upon the British Army.

In February 1900 Lord Roberts took over command and soon raised the sieges of Ladysmith and Kimberley and forced Cronje to surrender. By mid-June, after delays caused by shortages of supplies, Roberts had occupied Johannesburg and Pretoria. Mafeking was relieved. Buller advanced from Natal to join Roberts and Kruger fled.

Between October 1900 and May 1902, Kitchener was left to carry out the immensely difficult task of rounding up the Boer guerillas, a job made more difficult by the shortage of mounted troops. The war was brought to a bitter end by the use of a system of block-houses, farm burning and internment camps. The Transvaal and the Orange Free State became British colonies again and so remained till 1907.

137. The Boer War was by far the biggest of Victoria's reign and the army in the end deployed half a million men. Many of these were hastily raised volunteers and in this, as in other things, the Boer War foreshadowed greater wars to come. Shocked by the initial British disasters, the white dominions of the Empire sent contingents to South Africa. This photograph shows Canadian troops storming a kopje.

138. The price of war—wounded soldiers in a wagon shed at Klipdrift, 1900.

139. Although not quite on the scale of Mafeking, the defence of Ladysmith assumed symbolic significance in the minds of the British public. Here, the Gordon Highlanders line the route as the relief column enters the town.

140. Men of the 2nd Battalion, Grenadier Guards, filling water-bottles, 1901.

141. Clearing operations in the Brandwater basin.

SOURCES AND ACKNOWLEDGEMENTS

I am most grateful to the following for their kindness in lending illustrations: Mrs D.I. Clammer, 77; Dorset Military Museum (Curator: Lt Col D.V.W. Wakeley MC), 12, 18, 49, 72; The Institution of Royal Engineers, 109–13; Longsands Museum (Curator: G. T. Rudd), 22, 70, 76, 78, 128–36; B. Mollo, 71; National Army Museum, 1, 2, 3, 5, 6, 8, 9, 11, 14, 15, 16, 20, 23–7, 29, 30, 40, 54, 55, 64, 67, 68, 73–5, 83, 85, 86, 88, 91, 93, 95–108, 114–27, 137–41; Navy and Army Illustrated, 4, 7, 13, 17, 33–5, 37–9, 42–4, 51, 59–62, 65, 66, 69, 90; Radio Times Hulton Picture Library, 10, 19, 28, 47, 48, 56–8, 63, 89, 94; 13th/18th Royal Hussars (Lt Col D.A.G. Edelston), 21, 36, 53; 9th/12th Royal Lancers (Capt R.C. Peaper), 41, 46, 50, 52, 79–81, 84, 87; Royal Norfolk Regimental Museum (Curator: Lt Col A. Joanny MBE), 31, 32; Royal Sussex Regimental Association (Maj J.F. Ainsworth), 45, 92; Lt Col M.L.D. Skewes-Cox, 82.

For the kind help they have given me my especial thanks are due to: Mr B. Mollo, Keeper of Records, National Army Museum, and the staff of the Reading Room; Mr Y.W. Carman, also of the National Army Museum; Maj R.G. Bartelot, Royal Artillery Institution; Lt Col J.R. Palmer MC, 13th/18th Royal Hussars; Maj C.R.D'I. Kenworthy, Gordon Highlanders, who supplied me with information, and Mrs Sylvia Gelder, for typing the manuscript.

INDEX

Abyssinian campaign, 75-7
Addiscombe, 15
Afghan War, Second, 55, 81-3
Aldershot, 31, 41
Ali Musjid, 82
ammunition mule, 49
Annesley Bay, 74
Armstrong guns, 17
artillery practice, 42

Baden-Powell, R.S.S., 14
badminton, 61
Balaclava, 66
balloon, 54
band, regimental, 63
barrack-room law, 63
bayonet drill, 44
bicycle race, 61
big game, 57
Boer War, 19-20, 52, 54, 56, 64, 93-5
boxing, 65
Brandwater Basin, 95
bridge laying, 42
Brown, Sir G., 22
Buller, H.R., 13

Cambridge, Duke of, 11
Campbell, C., 10
Canadian troops, 93
carbine practice, 38, 39
Cawnpore, 69
Chard, Lt, 79
China War, Third, 72-4
Chitral Relief Expedition, 18, 84-6
Colt gun, 19
cricket, 60
Crimean War, 15, 22, 23, 24, 66-8
cyclists, 52

doolie, 52
Dorsetshire Regiment, 38, 88
Dragoon Guards, 5th, 24

East Lancashire Regiment, 49
elephants, 51

farriers, 32, 47
fencing, 44
field artillery team, 45
field hospital, 83
57th Foot (Middlesex Regiment), 67

Gatling gun, 17
Gordon, C.G., 12
Gordon Highlanders, 46
Gordon's palace, Khartoum, 92
Grenadier Guards, 43, 94
gunboat *Thafir*, 90
guns
  Armstrong, 17
  Colt, 19
  Gatling, 17
  Maxim, 18-19
  mortars, 15
  mountain, 16, 85
  pom-pom, 20
  railway, 20
  seige, 15
  12-pounder, 16

hanging mutineers, 71
heliograph, 55
Himalaya, HMS, 53
Hussars
  3rd King's Own, 31
  11th, 24
  13th, 31, 41

Imperial Yeomanry, 64
Indian Mutiny, 69-71
Irish Guards, 35
Isandhlwana, 78

Kafir-Bur Gate, Magdala, 77
Karachi gymkhana, 61
Khyber Pass, 82
King's Own Scottish Borderers, 87
King's Royal Rifles, 18-19
Kitchener, H.H., 13
Klipdrift, 93

Ladysmith
  balloon, 54
  relief of, 94
Lancers
  9th, 33
  12th, 33, 36, 40, 60
  17th, 32, 39, 47
  in training, 38-40
Life Guards, 32
Light Brigade, survivors, 68
Light Dragoons, 13th, 68
Lockhart, Lt-Gen, 89
Lucknow, 70, 71
luncheon party, 62

Madras Fusiliers, 1st, 70
Maxim guns, 18-19
Military Train, 48
mop-fighting, 64
mortars, 15
mountain artillery, 16, 50, 85
mounted infantry, 46

Napier, R.C., 10, 76
Nicholson, General, 89
North West Frontier, 16, 82

Omdurman, battle of, 91
Ootacamund, 62
oxen, 50, 51

Peh-tang fort, 74
Peshawar, 38
Peshawar Vale Hunt, 59
pig-sticking, 58
polo, 58
pom-pom, 20
Port Elizabeth, 26, 37
Prince Imperial, funeral parade, 80

Queen's Westminsters, 27

railways, 53, 90
railway gun, 20
Redan, 68
Rifle Brigade, 34, 65
Roberts, F.S., 11
rockets, Hale's, 21
Rorke's Drift, 79
route-marching, 37
Royal Bengal Fusiliers, 101st, 25
Royal Engineers, 42, 76, 79
Royal Horse Artillery, 34
Royal Irish Rifles, 86th, 26
Royal Norfolk Regiment, 28, 29
Royal Scots Fusiliers, 30
Royal Sussex Regiment, 35th, 36, 64

saddlery maintenance, 47
Seaforth Highlanders, 29, 30
seige guns, 15
semaphore, 56
Sevastopol, 56
Sherpur cantonment, Kabul, 82
shoeing-smiths, 48
sightseeing, 65
67th Foot (Hampshire Regiment), 81
68th Foot (Durham Light Infantry, 67
square, 38
Sudan, reconquest of, 90-2
sword-sharpening, 21

Taku forts, 73
telegraph, 55
tent-pegging, 59
theatricals, 62
Tirah campaign, 21, 52, 55, 56, 87-9
trench-digging, 37

Udney, Sir R., 89
Undul Wells, 76

village-burning, 88
volley-firing, 43

war correspondents, 56
Warwickshire Yeomanry Cavalry, 26
Wolseley, G.J., 12
wounded, 88, 93

Zula base camp, 75
Zulu War, 15, 78-80